P9-DNW-236

LEAPING GRASSHOPPERS

by Christine Zuchora-Walske

Lerner Publications Company • Minneapolis

The publisher gratefully acknowledges the assistance of Dr. Dan Johnson, Senior Research Scientist, Agriculture and Agri-Food Canada.

This book is available in two editions:
Library binding by Lerner Publications Company
Soft cover by First Avenue Editions
Divisions of Lerner Publishing Group
241 First Avenue North, Minneapolis, MN 55401 U.S.A.

Website address: www.lernerbooks.com

Words in *italic type* are explained in a glossary on page 30.

Library of Congress Cataloging-in-Publication Data

Zuchora-Walske, Christine
     Leaping grasshoppers / by Christine Zuchora-Walske.
         p.   cm. — (Pull ahead books)
     Includes index.
     Summary: Introduces the physical characteristics and behavior of grasshoppers.
     ISBN 0-8225-3634-X (hardcover: alk. paper)
     ISBN 0-8255-3638-2 (pbk. : alk. paper)
     .1. Grasshoppers—Juvenile literature.
     [1. Grasshoppers.] I. Title. II. Series.
     QL508.A2Z83 2000
     595.7'26—dc21   99-32633

Manufactured in the United States of America
1 2 3 4 5 6 — JR — 05 04 03 02 01 00

Look!
What is leaping by so quickly?

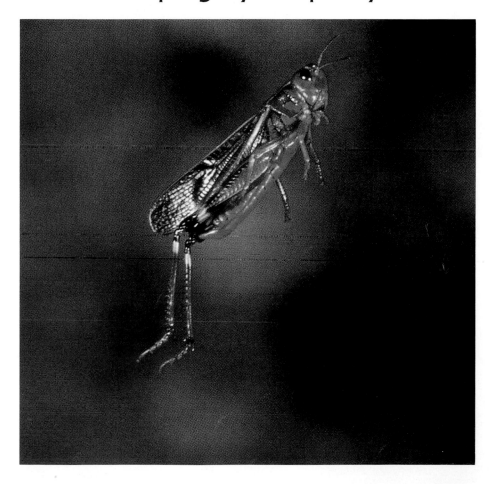

This is a
grasshopper.

A grasshopper is a kind of
animal called an *insect*.

An insect has three main
body parts and six legs.

Which legs are the longest
on this grasshopper?

A grasshopper has long,
strong back legs.

They help the grasshopper
leap fast and far.

They feel around with their
two *antennae.*

The long, skinny antennae stick
out of a grasshopper's head.

Grasshoppers also find food
with their five eyes.

The two big eyes can
see things moving.

Can you find three small eyes
between the two big eyes?

The small
eyes can
see only
light and
darkness.

A grasshopper's eyes, mouth, and antennae are all on its head.

The head is the front part of its body.

The middle part of its body
is called the *thorax.*

The back part of its body is
called the *abdomen.*

What do you see on the thorax of this grasshopper?

The legs are on the thorax.

Many grasshoppers have wings on their thorax, too.

Most grasshoppers with wings can fly, but they usually leap anyway.

Male grasshoppers can rub
their wings to make noise.

Why do they make noise?

Males make noise so females
can find them.

The females are looking
for male partners.

After a female finds a partner,
she will lay eggs.

This female grasshopper is laying
eggs underground.

She digs a hole with the back part of her abdomen.

Then she lays her eggs in the hole. She covers them with dirt.

Baby grasshoppers *hatch* from the eggs in the spring.

The babies are hungry!
They eat a lot and grow fast.

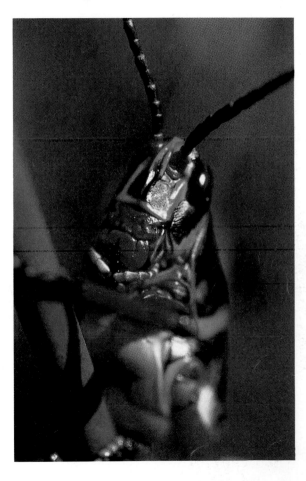

As they
grow, their
hard outer
shell gets
tight.

Growing grasshoppers must *molt*.

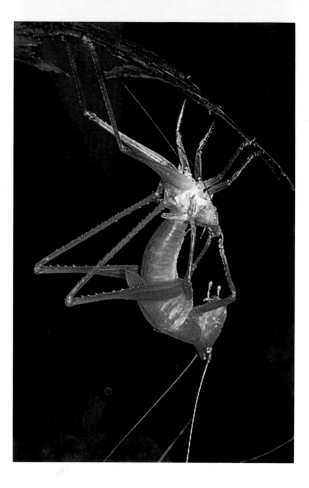

To molt, a grasshopper gets rid of its old, tight shell.

Take a walk on a warm, sunny day,

and watch the grasshoppers leaping away!

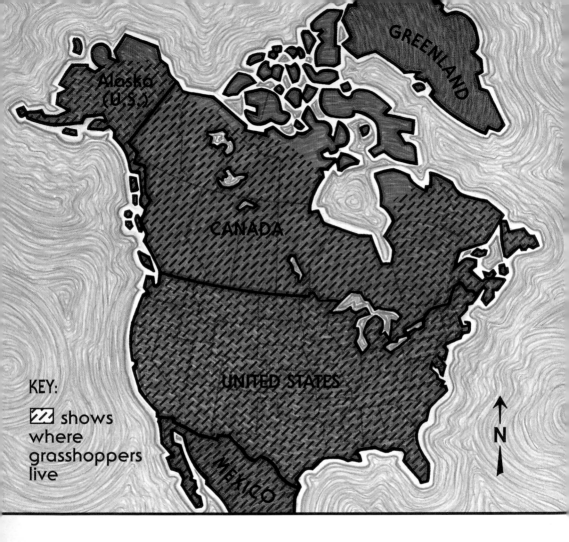

KEY:

▨ shows where grasshoppers live

Find your state or province on this map.
Do grasshoppers live near you?

Parts of a Grasshopper's Body

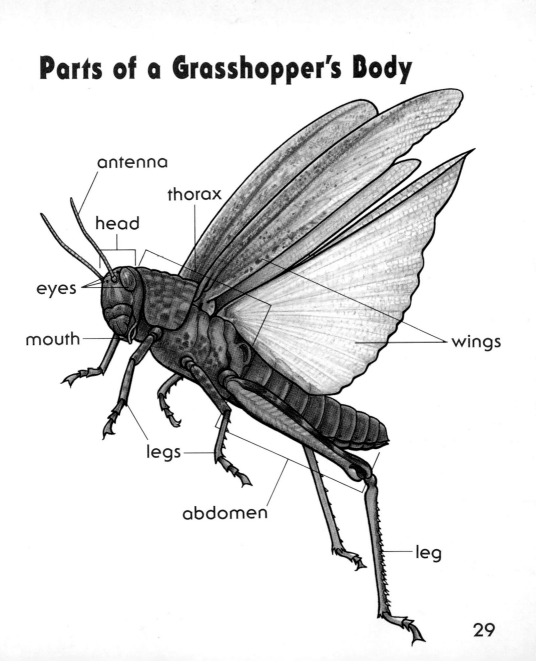

antenna

head

thorax

eyes

mouth

wings

legs

abdomen

leg

29

Glossary

abdomen: the back part of an insect's body

antennae: long, skinny feelers that stick out of a grasshopper's head

hatch: come out

insect: an animal with three main body parts and six legs

molt: get rid of an old, tight outer shell

predators: animals that hunt and eat other animals

thorax: the middle part of an insect's body

Hunt and Find

The publisher wishes to extend special thanks to our **series consultant, Sharyn Fenwick.** An elementary science-math specialist, Mrs. Fenwick was the recipient of the National Science Teachers Association 1991 Distinguished Teaching Award. In 1992, representing the state of Minnesota at the elementary level, she received the Presidential Award for Excellence in Math and Science Teaching.

Ron Zuchora-Walske

About the Author

Christine Zuchora-Walske grew up in Minnesota. She lived in Illinois for five years and loved to go hiking in the prairie, where there were lots of grasshoppers leaping every which way. She started to learn a lot about insects when she met her husband, who is a bug expert. Christine enjoys doing anything outdoors. She also enjoys making music and reading, writing, and editing books. Christine wrote *Peeking Prairie Dogs* and *Giant Octopuses* for Lerner's Pull Ahead series. She lives in Minneapolis with her husband, Ron.

Photo Acknowledgments

The photographs in this book are reproduced through the courtesy of: Visuals Unlimited: (© Gary Meszaros) front cover, p. 15, (© Rob & Ann Simpson) p. 4, (© John D. Cunningham) p. 14, (© William J. Weber) p. 26 (top), (© Richard L. Carlton) p. 27; © Robert & Linda Mitchell, back cover, pp. 7, 10, 16, 19; The National Audubon Society Collection/ Photo Researchers, Inc.: (© Hervy/ Jacana) p. 3, (© N. Smythe) p. 11, (© Holt Studios International/Nigel Cattlin) p. 12, (© Syd Greenberg) p. 13, (© Stephen Dalton) p. 17, (© J. H. Robinson) p. 18, (© L. West) p. 20, (© Gianni Tortoli) p. 21, (© David T. Roberts/ Nature's Images Inc.) p. 24; © Charles W. Melton, p. 5; © Bill Beatty, pp. 6, 8, 25, 26 (bottom left), 31; © Gary Braasch, pp. 9, 22, 23, 26 (bottom right).

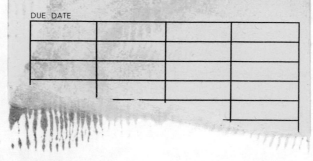

DUE DATE
